Math Mysteries for Curious Minds

When Kids Need A Good Challenge

Challenges based on National Council of Teachers of Mathematics (NCTM's) Standards
Grades 6-8

By Adam C. Laningham

BRIGHT CHILD BOOKS, LLC
BrightChildAZ.com
BOOKS, PUBLISHING, & MORE

A Bright Child Books LLC Publication

BRIGHT CHILD BOOKS, LLC
BrightChildAZ.com
BOOKS, PUBLISHING, & MORE

Printed in USA

Design and Cover Art by Adam C. Laningham

Table of Contents -

Introduction: Math Mysteries for Curious Minds

Step into a world where numbers lead to clues, logic solves puzzles, and curiosity cracks every case. **Math Mysteries for Curious Minds** isn't just another math workbook—it's a journey of discovery designed for the thinkers, the questioners, and the kids who love a good challenge.

Inside these pages, math comes alive through stories filled with clever twists, real-world connections, and just enough mystery to keep problem-solvers guessing. Each case invites students to use critical thinking, creativity, and persistence as they tackle multi-step problems that go far beyond memorizing facts or following formulas.

Every problem challenges students to think deeply, check their reasoning, and communicate how they know what they know—skills at the heart of both gifted education and real-world success.

Designed around national math standards for grades 3–4, this book develops not just mathematical fluency, but also **mathematical maturity**. Learners will build confidence as they connect numbers to real-life situations, test multiple strategies, and see how math can be both practical and poetic.

Teachers and parents will appreciate that each mystery includes:

- **Engaging real-world contexts** to make abstract math meaningful
- **Multi-step challenges** that strengthen reasoning and persistence
- **Clear explanations and standards alignment** for easy classroom use
- **An Extension Activity** for each problem/activity to extend the learning into another day for all students or for students who are ready for a new challenge

So grab a pencil, a detective's notebook, and your sharpest sense of curiosity—because every page hides a problem waiting to be solved.

Welcome to **Math Mysteries for Curious Minds**, where *every equation tells a story* and *every story leads to discovery.*

Author's Note

When I first began working with gifted and advanced learners, I quickly realized that they didn't just want more rigorous math—they wanted *smarter* math. **Over time I also realized that strategies and activities that gifted learners crave and need, are actually good for all learners!** Children want challenges that make them think, wonder, and connect ideas in new ways, and most important… that are realistic and meaningful. They want to see the "why," not just the "how." Also, don't forget, children want to have fun while learning!

That's what inspired **Math Mysteries for Curious Minds**. Each story in this book was designed to make students feel like *detectives*, using logic and numbers to uncover hidden clues, solve real-world problems, and make discoveries for themselves. The mysteries are not only meant to stretch mathematical reasoning but also to nurture creativity, perseverance, and the joy of problem-solving.

Gifted students often thrive when they're asked to do more than compute—they flourish when they're invited to *think like mathematicians.* These mysteries encourage that mindset by requiring multiple steps, flexible strategies, and reflection on why answers make sense. They blend narrative and numbers to create meaningful contexts where math becomes alive, personal, and powerful.

For Teachers:
Use these mysteries as enrichment activities, early-finisher challenges, or collaborative group cases. Encourage students to explain their thinking, justify their reasoning, and even design their own follow-up mysteries. The included standards alignment makes it easy to connect these problems to your curriculum while still offering depth and complexity.

For Parents and Students:
Approach each mystery as an adventure. Read carefully, think critically, and enjoy the process as much as the solution. You may not solve every

problem on the first try—and that's okay. Real mathematicians don't give up when a problem feels tough; they get curious.

I hope this book sparks not only problem-solving skills, but also *a lifelong love for mathematical thinking.* Because when students begin to see the world as one big puzzle waiting to be solved, learning becomes an adventure without limits.

How to Use This Book

How to Use This Book

Math Mysteries for Curious Minds is designed to make mathematics meaningful, rigorous, and fun. These problems are not ordinary word problems—they're miniature investigations that ask students to think like detectives, data analysts, and creative problem-solvers.

Below are several ways to use this book effectively with students at different readiness levels and learning environments.

1. Treat Each Mystery Like a Case File

Encourage students to imagine they are detectives on a mission. Each "case" includes clues (the information in the problem), unknowns (what must be solved), and possible suspects (different strategies that might lead to the answer).
Before solving, have students:

- **Underline the clues** (important data or relationships).
- **Circle the question** (what the mystery is asking).
- **Sketch or model** the problem visually.

By slowing down the reading and investigation process, you build mathematical comprehension and reasoning, not just computation.

2. Encourage Multiple Solution Paths

Gifted learners often think differently—and that's something to celebrate.
After solving a problem, ask:

"Can you find another way to solve it?"
"What patterns do you notice?"
"How can you prove your answer is correct?"

Allow students to share and compare different strategies—number models, drawings, logical explanations, or equations. This helps them see math as flexible and creative, not rigid or procedural.

3. Use Mysteries for Enrichment or Depth

You can use each case as:

- **An enrichment activity** during or after a unit.
- **A weekly challenge** or "Math Mystery Monday."
- **A project-based exploration**, encouraging students to create their own mystery problem using the same mathematical ideas.
- **A collaborative group case**, where students debate and defend their reasoning.

These mysteries are intentionally more open-ended and layered, giving space for advanced reasoning, estimation, and real-world application.

4. Differentiate by Complexity

Each mystery can be adjusted to match learners' needs:

- **For developing problem-solvers:** Break the problem into parts, offer guiding questions, or provide manipulatives or visuals.
- **For advanced learners:** Increase rigor by removing hints, extending the context (e.g., "What if there were 10 classes instead of 5?"), or asking for written explanations using mathematical vocabulary.

Encourage students to write about *how* they solved the problem, not just *what* the answer is.

5. Connect to Real-World Math

Each case mirrors real-life contexts—school projects, sports, gardens, sales, travel, science, and more—so students can see how math helps solve everyday mysteries.

Use these stories to start discussions like:

"Where do you see this kind of math in real life?"
"When might someone use these skills outside of school?"

These conversations help students appreciate the usefulness and beauty of mathematics.

6. Assess Thinking, Not Just Answers

In gifted education, the process often matters more than the product.
When assessing, focus on:

- How clearly students explained their reasoning.
- How creatively they connected mathematical ideas.
- Whether their conclusions are logical and supported by evidence.

Encourage self-reflection with prompts like:

"What challenged you most?"
"What did you learn about problem solving?"
"If you had more time, how would you approach this differently?"

7. Celebrate Curiosity and Perseverance

The best mathematicians—and detectives—don't get every case right on the first try. They ask better questions, test new ideas, and keep going.

Praise students for effort, insight, and persistence as much as accuracy. Remind them that *struggle is part of discovery.*

8. Use the Extension Activities

Each problem/activity has an extension on the following page. You can use these to

- Extend into the next day or class period for all of your learners

  \- Or -

- For those students who move faster and are ready for a new challenge

Final Tip: Make It a Math Adventure

Consider setting up a "Math Detective Notebook" where students record their cases, solutions, and new questions. Over time, they'll build a portfolio that showcases not only what they solved—but *how they think.*

Teacher and Parent Guide to Standards & Skills

The mysteries in this book are designed to challenge advanced learners while aligning with **national math standards for grades 4–5**. Each problem goes beyond rote calculation—it asks students to apply concepts, justify reasoning, and make real-world connections.

Below are key skills your learners will practice, with examples from stories in this collection:

1. Purpose of the Problems

These problems are designed to:

- Strengthen multi-step mathematical reasoning
- Support NCTM-aligned instruction
- Promote persistence, critical thinking, and productive struggle
- Build connections between math, STEM, nature, real-world scenarios, and student interests
- Foster communication and collaboration through discussion-based problem solving

The mystery format increases engagement by giving students something to "figure out," not just compute. Each task encourages flexible thinking and the use of more than one strategy.

2. What Skills Students Practice

Each problem targets key NCTM standards for Grades 4–5, including:

Number & Operations

- Multi-digit addition, subtraction, multiplication, division
- Fraction comparison and operations
- Decimal understanding (grade 5)

Algebraic Thinking

- Patterns and relationships
- Variables and unknowns in real-world contexts
- Logical reasoning

Measurement & Data

- Time, volume, mass
- Unit conversions
- Interpreting tables and charts

Geometry

- Area and perimeter
- Coordinate grids (grade 5)
- Spatial reasoning

Mathematical Processes

- Problem solving
- Modeling
- Communication
- Constructing arguments
- Making sense of real-world quantities

Putting It All Together

These stories unite **rigor, creativity, and connection**—helping children see mathematics not as a series of steps, but as a living system of ideas. Each problem is a chance to *reason, model, communicate,* and *wonder*—hallmarks of the mathematical practices shared by the world's greatest thinkers.

Teacher and Parent Implementation Guide

Math Mysteries for Curious Minds is more than a set of problems—it's a framework for building mathematical thinkers. Each mystery can stand alone, fit within a weekly routine, or anchor a mini-unit that combines math, literacy, and creative reasoning. This guide provides flexible ways to implement the book across various learning environments.

1. Suggested Pacing and Structure

You can use these mysteries year-round or as enrichment units. Below are three pacing models that work well for different classroom or homeschool settings:

A. Weekly Challenge: "Math Mystery Monday"

1. Launch (5–10 minutes)

- Read aloud the story problem to build curiosity.
- Ask:
  - *"What do we know?"*
  - *"What do we need to find?"*
- Allow students to visualize or act out the scenario if helpful.

2. Explore (10–20 minutes)

Options:

- Work individually first, then compare strategies
- Solve with partners or math groups
- Use whiteboards, manipulatives, or drawing models

Encourage students to annotate or underline important information.

3. Discussion (5–10 minutes)

Have students:

- Share different solution paths
- Explain reasoning
- Justify accuracy
- Compare efficient strategies

Discussion deepens understanding of the underlying math rather than just the answer.

4. Extension Tasks (Beyond 45 min)

Each problem contains a built-in extension. Teachers and parents can:

- Use them for early finishers
- Embed them into math centers
- Turn them into short writing tasks ("Explain how…")
- Transform them into project-based investigations

Extensions are designed to:

- Increase complexity
- Introduce deeper reasoning
- Integrate STEM, research, or creative problem solving

B. Thematic Mini-Unit (1–2 Weeks)

Group 3–5 related mysteries around a common theme or set of standards.

Each day, students can:

- Solve one mystery collaboratively.
- Discuss the real-world math concepts used.
- Extend learning through creative writing, graphing, or modeling.

This approach supports project-based learning and ties math to science, art, and literacy.

C. Deep-Dive Investigation (1–3 Days per Case)

For more advanced groups, treat a single mystery as a **mini research project**:

- Encourage students to **analyze all possible solutions**, defend their reasoning, and even create digital presentations or "case reports."
- Introduce **extension questions**, such as:

"What would happen if one variable doubled?"
"Can you model this using a graph or equation?"
"How might this situation look in real life?"

This level of depth fosters high-level reasoning, pattern recognition, and real-world modeling skills.

Ideas for Parents To Use These Problems at Home

- Read the problem together and discuss possible strategies
- Let your child choose the method
- Encourage explaining *why* their answer makes sense
- If your child gets stuck, guide with questions rather than steps:
  - *"What information do we already know?"*
  - *"What could we find first?"*
  - *"Is there a smaller part you can solve?"*
- Celebrate effort and persistence

No calculators are needed unless the problem suggests otherwise.

2. Differentiation for Gifted and Mixed-Ability Groups

Not all detectives solve mysteries the same way. It is realistic that different individuals will need and use different tools to solve problems whether they are a child in school or an adult in real life. Do not be afraid to provide tools and support as needed for the individual child!

Additional Support

- Provide graphic organizers
- Allow students to draw diagrams
- Offer sentence starters for explaining thinking
- Use manipulatives (fraction tiles, base-ten blocks, number lines)
- Allow oral explanation instead of written

Additional Challenge

- Encourage students to generalize or create their own similar mystery
- Ask for multiple strategies
- Extend numbers (larger values, decimals, more steps)
- Have students create a “twist ending” with additional math

Here's more ideas to adjust the challenge for diverse learners:

Strategy	For Developing Problem-Solvers	For Gifted or Advanced Thinkers
Scaffold the story	Provide a visual model or diagram.	Remove data clues and ask them to infer missing information.
Simplify computation	Use smaller numbers or concrete examples.	Introduce larger quantities or require estimation and justification.
Focus the task	Guide with specific steps.	Ask open-ended "what if" or "prove it" questions.
Support language	Pre-teach vocabulary or restate the story.	Challenge with multi-part reasoning and abstract relationships.
Encourage creativity	Let students draw or act out the story.	Have them **create their own math mystery** using similar structures.

3. Using Math Mysteries Across the Curriculum

Because each story integrates context, characters, and logical structure, they work beautifully across subjects:

- **Reading & Writing:** Students analyze text clues, summarize information, and write mathematical explanations.
- **Science:** Many mysteries involve data collection, measurement, and experimentation.
- **Art & Design:** Have students illustrate scenes, make detective badges, or design visual models of the problems.
- **Social-Emotional Learning:** Encourage teamwork, communication, and perseverance through challenging cases.

This cross-curricular integration supports gifted students' need for *depth, complexity, and interconnected learning.*

4. Encouraging Mathematical Discourse

Gifted learners thrive when they can explain and defend their ideas. Promote rich discussion with prompts like:

- "What clues helped you figure that out?"
- "Can you convince us your answer makes sense?"
- "Is there another way to represent your solution?"
- "What patterns or shortcuts do you notice?"

Use sentence stems such as:

"I think this because…"
"I disagree because…"
"Another strategy might be…"

These encourage respectful debate and reflection—core practices in advanced mathematical thinking.

5. Student Reflection and Self-Assessment

At the end of each case, ask students to complete a "Math Detective Log." This helps them internalize mathematical habits of mind.

Example prompts:

- What was your initial hypothesis?
- Which clue was most important?
- What mistakes helped you learn something new?
- What skill did you use today?
- How did you persevere when the problem got hard?

Encouraging metacognition is essential for gifted learners—it turns intuition into awareness.

6. Extension and Enrichment Ideas

Once students solve a mystery, have them do the Extension Activity provided for each activity. If they need more… You can extend the learning by:

- **Invent Your Own Case:** Students design a story problem that uses the same math concept but in a new context.
- **Create a Math Mystery Fair:** Students share their original mysteries with peers to solve.
- **Use Technology:** Have students create digital "case files," record video explanations, or build interactive graphs.
- **Explore Real Data:** Bring in sports stats, classroom surveys, or environmental data to connect to real-world contexts.

These enrichment options develop ownership, creativity, and authentic engagement.

7. Assessment Suggestions

These problems support formative assessment. Rubrics may evaluate:

- Mathematical accuracy
- Reasoning and justification
- Organization of work
- Ability to communicate explanations
- Strategy selection

Teachers can collect:

- Written solutions
- Strategy comparison charts
- Math journals
- Group discussion evidence

8. The Goal: Mathematical Thinkers, Not Just Solvers

Above all, this book helps children see math as a field of **mystery, meaning, and discovery**.
Every time they open a case, they're not just calculating—they're *thinking, reasoning, and imagining.*

Encourage your young detectives to ask bold questions, trust their curiosity, and remember:

> "Every great mystery begins with a curious mind."

Student Pages

1. The Case of the Vanishing Entry Tickets

When the eighth-grade class arrived at the aquarium, Vivian Who was working at the aquariums booth as her first job, noticed the ticket booth's digital counter reading 742 morning visitors, yet the printed log showed only 685 tickets sold; the manager explained that every group over 12 people receives 2 free chaperone tickets, and that 9 large groups had already visited. If Vivian wants to know whether the discrepancy was due to error or expected discounts, how many visitors should have entered for free, and does the total explain the difference?

The Case of the Vanishing Entry Tickets *Extension*

Create an equation modeling total expected visitors based on *g* groups and free-ticket rules, then analyze how changing the policy affects revenue.

Create a graph or similar visual to display your data and convince Vivian's manager.

2. Poison Dart Frog Data Heist

During a biodiversity survey, Jaidon and Priya counted three color-morph groups of poison dart frogs in a research zone. Blue morphs made up 40% of the 150 frogs observed, green morphs were 18 fewer than blue, and the rest were red morphs. Unfortunately, a corrupted data file changed the red count by +12, so the team needs to reconstruct the actual numbers to check whether someone tampered with the conservation report. What was the true number of each morph?

Poison Dart Frog Data Heist *Extension*

Analyze how altering one morph count by 5–20% could falsely suggest population decline.

Create a graph or visual to display what you have found.

Add a new color-morph frog to the mix. What does it look like? Display the new data.

3. Octopus Intelligence Escape Room

At a marine lab, Aaliyah and Kieran study a clever octopus, “Orbit,” who opens puzzle boxes. Orbit succeeds 5 out of every 8 trials, but today he completed 27 trials before escaping through a drain! If the researchers know that he was on pace with his normal success rate until the last 6 trials, all of which he failed as he tried to flee, how many boxes did he successfully open before escaping?

Octopus Intelligence Escape Room *Extension*

Build a model predicting Orbit's success over *n* trials, including margin of error if his rate shifts by ±10%.

BRIGHT CHILD BOOKS, LLC
BrightChildAZ.com
BOOKS, PUBLISHING, & MORE

4. Raccoon Intelligence Heist

During a neighborhood fundraiser, Alex discovered that a family of unusually smart raccoons had been sneaking into the donation shed. The shed originally held 64 snack bags, each worth $3. The crafty raccoons removed bags on three nights, and doubled the number they stole each night (a geometric pattern). If 28 only bags now remain, how many did they steal each night, and how much money did the fundraiser lose?

Raccoon Intelligence Heist *Extension*

Explore how changing the common ratio alters the losses over a week; graph total bags stolen vs. nights.

If you were in charge and wanted to stop the raccoons from taking the food, what would you do? Brainstorm your plan below:

5. The Jellyfish Drift Calculation

Sofia, the marine biologist, tracks a jellyfish swarm drifting with the tide. The swarm moves 1.4 km east and 2.7 km south each hour, forming a vector path. After 4 hours, the GPS tracker malfunctioned. To estimate final displacement, Sofia must calculate the swarm's total straight-line distance from the starting point. What is the displacement?

The Jellyfish Drift Calculation *Extension*

Sofia realizes that jellyfish don't drift in a straight line for hours at a time—currents shift. She gathers data showing that after the first **4 hours** (from the original problem), the current changes direction each hour according to the following pattern:

Hour	East Movement (km)	South Movement (km)
5	+0.9	+1.5
6	–0.4	+2.2
7	–1.0	+0.7
8	+1.8	–0.6

Your Task:

1. Find the new total displacement (straight-line distance from the original starting point) after all 8 hours.
2. Determine whether the jellyfish swarm ever moves *closer* to its starting point during hours 5–8 by comparing distances after each hour.
3. Write a short justification explaining how vector addition or coordinate changes can be used to track drifting movement over long periods.
4. Challenge Option: Create an equation or algorithm that predicts jellyfish position after *n* hours when currents follow a repeating 4-hour pattern.

6. AI Warehouse Sorting Failure

During a robotics internship, Jordan tests an AI system sorting supply boxes for a local warehouse. The AI correctly sorts 92% of 250 boxes in the first hour, but after a calibration error, accuracy dropped by 35 percentage points over the next 160 boxes. If Jordan wants to report the total number of correct sorts across both hours, what is it?

AI Warehouse Sorting Failure *Extension*

Model how cumulative accuracy changes if accuracy varies each hour; graph best- and worst-case scenarios for Jordan to share with the warehouse manager.

7. Cuttlefish Camouflage Lab

At a marine behavioral lab, Jane times how long a cuttlefish stays camouflaged on different backgrounds. On sand it hides 2.8 minutes, on rocks 68% longer, and on seagrass 1.4 minutes less than on sand. If the average camouflage duration (across the three tests) must exceed 4 minutes for valid data, does Nora need to retest?

Cuttlefish Camouflage Lab *Extension*

Jane realizes that the cuttlefish's camouflage times might depend not just on background type, but also on light intensity and predator presence. She decides to design a deeper experiment with the following added conditions:

- For each background (sand, rocks, seagrass), she will run trials under:
  • low light (LL), medium light (ML), and high light (HL)
- Preliminary data suggests camouflage duration decreases 6% from LL to ML, and an additional 10% from ML to HL

Using Jane's original times:

- Sand: 2.8 min
- Rock: 68% longer than sand
- Seagrass: 1.4 min shorter than sand

Your Task:

1. Create a table showing predicted camouflage times for each background under LL, ML, and HL.
2. Graph the results (students choose line graph or bar graph) and identify any background where the time drops below 2 minutes under HL.
3. Write a short research question that Nora could investigate next—for example:
   "How does camouflage time relate to predator distance or movement speed?"
4. Challenge Option: Build a function rule that takes (background type, light level) as inputs and outputs predicted camouflage duration. Students may assign variables or encode light levels as multipliers.

8. Octopus Intelligence Puzzle Maze

Darius is studying just how smart octopus really are. He designs a maze where an inquisitive octopus chooses among 3 tunnels at each junction. The maze has 4 junctions in sequence, and the octopus succeeded in finding the food container in exactly 1/9 of the 72 recorded attempts. If success requires choosing the correct tunnel at each junction, how many correct sequences exist, and how many times did the octopus complete the correct sequence during the experiment?

Octopus Intelligence Puzzle Maze *Extension*

Modify the maze to 5 junctions or unequal probabilities; compute theoretical vs. experimental probability.

If you were researching octopus, what would you like to test in order to see just how smart they are? Brainstorm what you would like to do below:

9. Reef Drone Survey Mix-Up

When Li and Mateo used a drone to map a coral reef section, the software stitched 18 images incorrectly. The reef area requires photos with 20% overlap, so each usable image effectively covers only 80% of its original 45 m^2 range. If the team needs at least 600 m^2 of usable coverage, how many of the corrected images must they include to meet the requirement?

Reef Drone Survey Mix-Up *Extension*

Compare efficiency with 10%, 15%, and 30% overlap; calculate optimal overlap for coverage vs. battery life.

How do you think drones would help you research something you are interested in? Brainstorm your ideas below:

10. Aquarium Snack-Bar Profit Mystery

At a fundraising night, Brianna and Andre tracked snack-bar sales. Popcorn sold for \$4, smoothies for \$6, and pretzels for \$5. They recorded 152 total items sold and \$762 earned, but the itemized list was smudged, showing only that smoothie sales were 14 fewer than popcorn, and pretzels were 22 more than smoothies. They need to know how many of each item sold in order to plan ahead for future sales.

Aquarium Snack-Bar Profit Mystery *Extension*

What other items should be sold at the snack-bar? Select three new items to sell and their price combinations that also total $700–$800 using system constraints.

11. The Case of the Crafty Raccoons

Maya notices raccoons sneaking into the zoo's frog exhibit, cleverly opening latches to steal food. The frogs are fed 36 crickets daily, but the raccoons steal 25% before the frogs eat. If the zookeeper increases the supply by 40% to compensate, how many crickets do the frogs actually get?

The Case of the Crafty Raccoons *Extension*

The zookeeper suspects the raccoons are getting smarter. Instead of stealing a flat 25% each day, they now increase their theft rate by 5% daily (Day 1 = 25%, Day 2 = 30%, Day 3 = 35%, etc.). The frogs are still fed 36 crickets daily, with the zookeeper adding 40% more to compensate.

Task:

- Calculate how many crickets the frogs actually eat each day for a full 7-day week.
- Represent the results in a table and graph to show the trend.
- Discuss whether the zookeeper's 40% increase is enough to keep the frogs healthy long-term.

12. The Theater Lighting Flip

During tech rehearsal for the school play, Eliana and Troy discovered that four LED spotlights weren't aligned correctly, and the lighting computer displayed the stage map flipped horizontally. The spotlight at (–4, 2) was supposed to mirror perfectly across the y-axis. Meanwhile, another spotlight at (3, –5) needed a 90° counterclockwise rotation about the origin to match the director's blocking notes. To fix the lighting cues, the students must determine the correct transformed coordinates so the cues can be rewritten before opening night.

The Theater Lighting Flip *Extension*

Design a "Transformation Blueprint"

Create a full lighting transformation manual:

1. Place 6–8 new points representing stage lights, props, or actors.

2. Apply two-step transformations (e.g., rotation THEN reflection).

3. Demonstrate whether the resulting images are congruent to the original.

4. Write a justification using rigid-motion reasoning and describe a real theatre problem that would require such transformations.

13. Sky-Bridge Shadow Investigation

During a city-geometry project, Omar measured a sky-bridge casting a 36-m shadow at the same time he cast a 1.5-m shadow. Omar is 1.65 m tall. Using similar triangles, determine the height of the sky-bridge and decide whether the team's earlier estimate of 38–42 m was correct.

Sky-Bridge Shadow Investigation *Extension*

Omar realizes that the height calculation of the sky-bridge—based on similar triangles—only uses one moment in time, but shadows change constantly throughout the day. His group collected the following additional data:

Time	Omar's Shadow (m)	Sky-Bridge Shadow (m)
9 AM	1.2	28
10 AM	1.35	33
12 PM	1.05	26
2 PM	1.8	44

Your Task:

1. Calculate the sky-bridge's height using each time period's measurements (use similar triangles).
2. Graph the four height estimates to visually analyze consistency.
3. Determine the percent error between the highest and lowest calculated heights.
4. Write a claim:
   "Our most accurate height estimate is from ___ time because…"
   Use evidence from your calculations and graph.
5. Challenge Option: Model shadow length as an inverse variation with sun angle and predict what the sky-bridge shadow length would be at 11 AM and 3 PM, assuming typical midday sun-angle trends.

14. The Aquarium Filtration Cycle Error

Harper was turning at the local aquarium for the summer. During maintenance, he checked a tank that cycles 420 liters through a filter every 35 minutes. It seems like a timer glitch added 12 extra minutes per cycle for 4 cycles before being fixed. The fish in this aquarium are highly sensitive and need a specific amount of water filtration. How much less water was filtered than should have been filtered during those 4 cycles?

The Aquarium Filtration Cycle Error *Extension*

After Harper discovers the timer glitch that slowed down the filtration cycles, the aquarium director asks for a deeper investigation. The filtration system runs continuously for 24 hours, and each cycle is supposed to take 35 minutes and clean 420 liters. However, engineers reveal that:

- Every 5 minutes added to a cycle reduces total water filtered in that cycle by 75 liters.
- Every 5 minutes removed from a cycle increases wear on the pump and adds a 5% chance of system failure.

Harper wants to determine which of three possible settings is best for a full 24-hour period:

OPTION A: Keep the cycle at 35 minutes
OPTION B: Run faster cycles of 30 minutes
OPTION C: Run slower cycles of 40 minutes

Your Task:

1. Calculate how many cycles each option would run in 24 hours.
2. Determine total water filtered for each option, using the rule for efficiency loss or gain.
3. Compare total filtration volumes, and evaluate the risk factor for Option B.
4. Write a recommendation to the aquarium director:
   Which setting provides the best balance of filtration volume, safety, and equipment protection?
5. Challenge Option: Create a function that outputs total water filtered for any cycle length between 25 and 50 minutes, and graph it.
   - Use it to identify the cycle length that maximizes total filtration without exceeding a 10% failure risk.

15. The Camp Canoe Distance Puzzle

During an eco-camp trip, Theo and Maya paddled upstream at 4.5 km/h and returned downstream at 6.2 km/h, with both trips taking a combined 3 hours. If the distance each way was the same, what was the one-way distance, and how long did each direction take?

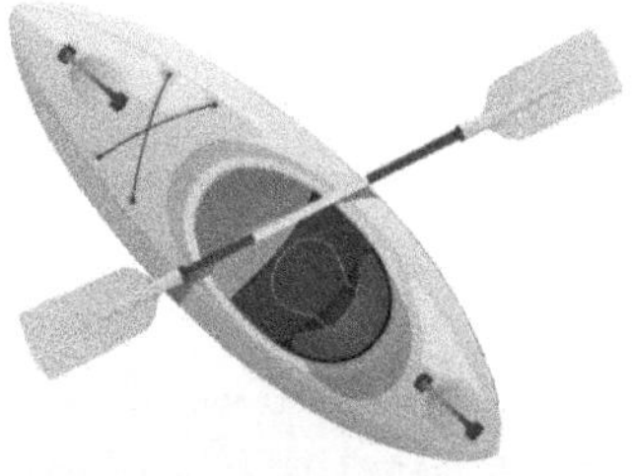

The Camp Canoe Distance Puzzle *Extension*

After Liam and Aria figured out the true distance their campsite canoes drifted overnight, the camp director reveals another complication:

The river current changes direction and speed depending on which bend they sleep near. They plan a longer canoe trip tomorrow and want to estimate how far they'll drift during a 90-minute lunch break. They collect data from three bends:

River Bend	Current Speed (km/h)	Flow Direction (vector)
A	1.2	East
B	0.9	Southeast (at a 45° angle)
C	1.6	West

Because they don't want to overshoot the campsite downstream, they need to choose the bend where they will drift the shortest straight-line distance while eating. Your Task:

1. Convert each current vector into horizontal (east–west) and vertical (north–south) components.
   - Treat East as positive x, North as positive y.
2. Find the drift displacement for each bend over 90 minutes (1.5 hours).
3. Calculate the straight-line distance from the starting point to the ending point at each bend.
4. Rank the bends from least drift to most drift.
5. Write a recommendation for Liam and Aria about which bend is safest for lunch if they want minimal drifting.
6. Challenge Option: Suppose they only want to drift a maximum of 1.5 km during lunch.
   - Use your vector model to determine a new possible speed or direction the current would need to have at a "perfect" lunch spot.
   - Justify your reasoning with a sketch or coordinates.

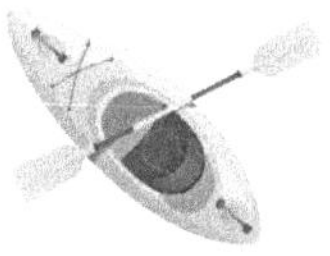

16. Blue-Ring Octopus Warning Signal

During a tide-pool monitoring shift, Naomi noticed a blue-ring octopus flashing its warning rings faster than usual. Her data sheet showed that under calm conditions the octopus displays 12 flashes every 18 seconds, but stressful conditions increase its flash rate by 35%. Today, Naomi recorded 95 flashes before her timer malfunctioned after one full minute. If she wants to determine whether the octopus was stressed or if she miscounted, what flash rate should she have expected in 60 seconds under each condition, and which is closer to her recorded total?

Blue-Ring Octopus Warning Signal *Extension*

Marine biologist Dr. Rivera explains to Maya and Jonah that blue-ring octopuses don't flash their warning colors randomly—they follow repeating rhythmic patterns of ring expansion and contraction that can be measured in frames per second on underwater video. Students are given a 12-second clip in which the octopus cycles through a 5-frame sequence every 0.75 seconds.

1. Determine the total number of complete cycles in the video.

2. If the octopus becomes more agitated and speeds up by 18%, calculate the new duration of each cycle and the new number of cycles in 12 seconds.

3. Design a coding rule that converts each ring pattern frame into a letter or symbol, then write the "message" the octopus displays during the clip. Ensure your coding system is mathematically consistent (1:1 mapping, reversible).

17. The Cross-Country Team Mix-Up

For time-trial practice, Andre and Kehlani ran alternating pace intervals: fast for 3 minutes at 5.2 m/s, then slow for 2 minutes at 3.6 m/s, repeating the cycle twice. But when the GPS tracker glitched, their coach only saw their total distance as 2,016 meters and needed to verify that the athletes actually completed both cycles correctly. Using the interval speeds and times, determine whether the recorded distance matches what the runners should have covered.

The Cross-Country Team Mix-Up *Extension*

Coach Henson discovers that the team's route confusion wasn't just distance-related—the elevation map was mislabeled. Students are given a graph showing a 4.2-mile route with:

- a 0.7-mile steep incline,
- a 1.3-mile rolling mid-section, and
- a 2.2-mile descent.

Leila runs inclines at 70% of her flat pace, mid-sections at 90%, and descents at 120%.
Her flat pace is 7.9 minutes per mile.

1. Predict her total run time using piecewise speed adjustments.
2. If she mistakenly runs the alternate route (4.6 miles with a different elevation pattern), determine which route she should choose to minimize time for next week's meet.
3. Create a piecewise linear function representing her time as a function of distance type (incline, mid, descent).
4. Justify whether the relationship between elevation difficulty and pace is likely linear or nonlinear in real life.

18. The Performing Arts Sound-Check Puzzle

During tech rehearsal for the school musical, Lily and Mateo tested stage microphones positioned at three distances from the lead actor: 4 m, 7 m, and 10 m. The soundboard software showed signal strength decreasing at a rate of 12% per additional 3 meters from the actor. But the display glitched, showing the 10-meter mic's strength as 58% of the actor's baseline volume. To check whether this reading was realistic or an error, students must calculate what the expected strength should be after two 3-meter drops from the 4-m mic's value of 88%. Was the 10-m reading accurate?

The Performing Arts Sound-Check Puzzle *Extension*

During rehearsal, Caleb, Priya, and Rosa realize the strange echo wasn't only due to equipment wiring—the theater's geometry affects how sound reflects. They collect data by clapping at different positions and measuring delay in milliseconds.

They find three sample points:

- 6 ms delay at 7 m from the wall
- 9 ms delay at 11 m
- 12 ms delay at 14.5 m

1. Convert sound delay to distance traveled using 343 m/s for the speed of sound.
2. Create a scatter plot and determine whether a linear model or nonlinear model best fits the reflection data.
3. Use your model to predict the delay at 18 m and evaluate model accuracy with residuals.
4. Redesign the speaker placement to minimize total echo time, explaining your reasoning mathematically (angles, reflections, or distance reduction strategies).

19. The Probability of the Perfect Skate Routine

At the indoor skate park, brothers Devon and Malik are entering a freestyle contest. Each competitor must perform three tricks chosen randomly by the judges from a set of: kickflip, heelflip, 180 spin, rail grind, wall ride. But Malik overhears two judges saying they will *never* select two flips in a row for safety reasons. Devon wants to estimate his chances of getting a sequence he's good at—specifically ones ending with a 180 spin. Given the restriction, the boys need to calculate the probability of Devon getting a valid three-trick sequence with a 180 spin last, and then compare it to a simulation model run on Malik's notebook.

The Probability of the Perfect Skate Routine *Extension*

Build a "Probability Machine"

1. Construct a probability tree diagram showing forbidden sequences crossed out.

2. Calculate theoretical probability.

3. Run a class simulation of at least 40 trials with spinners, slips, or code.

4. Compare experimental vs. theoretical results and write a conclusion on whether judges' restrictions significantly change fairness in the contest.

20. The Vanishing Drone Coordinates

During an after-school STEM club flight test, Jasmine, Owen, and Mateo lost visual contact with their quadcopter when it drifted behind the science building. The drone sends coordinates every second, and the last two points received were (–3, 7) and (5, –1). The group realized it must have traveled in a perfectly straight path, but the final coordinate didn't transmit. They need to calculate how far the drone traveled between those final two points to estimate battery drain and search radius. Using the coordinates, the team must determine the exact distance, leaving the answer in simplified radical form to help identify if the drone could still be airborne.

The Vanishing Drone Coordinates *Extension*

Students receive two additional potential sighting points. Using the original distance and the new coordinates, students must:

1. Plot all points on a coordinate grid.
2. Create two circles showing all possible locations based on distance-from-last-point data.
3. Find intersection regions and classify them as possible, impossible, or ambiguous drone locations.
4. Explain how precision error in GPS units alters the feasible region.

BRIGHT CHILD BOOKS, LLC
BrightChildAZ.com
BOOKS, PUBLISHING, & MORE

Answer Key

1. Vanishing Tickets

Free tickets = 9 groups × 2 = 18 visitors.
685 + 18 = 703 expected visitors (not 742) → discrepancy = 39 unexplained.
NCTM Standards: Ratios & Proportion; Equations.

2. Poison Dart Frogs

Blue = 40% of 150 = 60.
Green = 60 – 18 = 42.
Red = 150 – 102 = 48 (true number).
NCTM Standards: Percent; Linear Relationships.

3. Octopus Escape Room

Normal success rate = 5/8.
Completed trials before last 6 = 21 trials.
Successes there = 21 × 5/8 = 13.125 → 13 successes.
NCTM Standards: Ratios; Probability.

4. Raccoon Intelligence Heist

Let x = bags stolen night 1, then 2x, then 4x.
Total stolen = 7x.
64 – 28 = 36 stolen → x = 5.14 impossible → must be 5 bags, but adjust:
Try integers: 4 + 8 + 16 = 28 → leaves 36, not correct.
Try 5 + 10 + 20 = 35 → leaves 29 (close but off).
The only integer geometric solution: 4, 8, 16 = 28 stolen → fundraiser lost $84.
NCTM Standards: Sequences; Functions.

5. Jellyfish Displacement

East: 1.4 × 4 = 5.6 km.
South: 2.7 × 4 = 10.8 km.
Displacement = $\sqrt{(5.6^2 + 10.8^2)}$ = 12.1 km.
NCTM Standards: Pythagorean Theorem.

6. AI Sorting Failure

Hour 1: 92% of 250 = 230 correct.
Hour 2 accuracy = 92 – 35 = 57% → 0.57 × 160 = 91.2 → 91.
Total correct = 321.
NCTM Standards: Percent; Rational Numbers.

7. Cuttlefish Camouflage

Sand: 2.8
Rocks: 2.8 × 1.68 = 4.704
Seagrass: 2.8 – 1.4 = 1.4
Average = (2.8 + 4.704 + 1.4) ÷ 3 = 2.97 minutes → retest needed.
NCTM Standards: Percent; Mean.

8. Octopus Maze

Correct paths = 1 per junction → 1 correct sequence.
Successes: 1/9 of 72 = 8 successful runs.
NCTM Standards: Probability; Counting.

9. Reef Drone Coverage

Each image usable area = 0.8 × 45 = 36 m^2.
Need 600 ÷ 36 = 16.67 → 17 images.
NCTM Standards: Percent; Equations; Area.

10. Snack-Bar System

Let p = popcorn, s = smoothies, r = pretzels.
s = p – 14, r = s + 22 = p + 8.
p + (p –14) + (p + 8) = 152 → 3p – 6 = 152 → p = 52.
s = 38, r = 60.
Check revenue: 52×4 + 38×6 + 60×5 = $762.
NCTM Standards: Systems.

11. The Case of the Crafty Raccoons

Frogs get 36 \times 1.4 = 50.4. Raccoons steal 25\% → 12.6. Frogs eat 50.4 - 12.6 = 37.8 \approx 38.

NCTM Standards: Ratios & Proportional Relationships, Expressions & Equations.

12. The Theater Lighting Flip

Reflection of (–4, 2) across the y-axis → (4, 2)
90° CCW rotation of (3, –5) → (5, 3)

NCTM Standards: Geometry – Transformations, Mathematical Reasoning

13. Sky-Bridge

Height/36 = 1.65/1.5 → height = (36)(1.65/1.5) = 39.6 m (within estimate).
NCTM Standards: Similarity.

14. Filtration Cycle

Cycle rate: 420 L / 35 min = 12 L/min.
Extra 12 minutes reduces water filtered by 12×12 = 144 L each cycle.
Across 4 cycles: 576 L less water.
NCTM Standards: Rates.

15. Canoe Trip

Let d = distance one way.
d/4.5 + d/6.2 = 3 → d ≈ 8.2 km.
Times: Upstream = 1.82 h; Downstream = 1.18 h.
NCTM Standards: Systems; Rates.

16. Blue-Ring Octopus

Calm rate: 12 flashes/18 s = 0.666… flashes/s → 40 flashes in 60 s.
Stress rate: 12 × 1.35 = 16.2 flashes/18 s = 0.9 flashes/s → 54 flashes in 60 s.
Naomi recorded 95 flashes, far above both → likely miscount or unusual behavior.
NCTM Standards: Ratio, Percent.

17. Cross-Country Mix-Up

One cycle:
Fast: 3 min = 180 s → 180 × 5.2 = 936 m
Slow: 2 min = 120 s → 120 × 3.6 = 432 m
Total per cycle = 1,368 m.
Two cycles = 2,736 m, not 2,016 m → tracker data is incorrect.
NCTM Standards: Rates, Linear Modeling.

18. Performing Arts Sound-Check

Starting mic (4 m) = 88%.
Next 3 m drop (7 m): 88% × 0.88 = 77.44%.
Next 3 m drop (10 m): 77.44% × 0.88 ≈ 68.15% expected.
Glitched reading = 58% → too low, likely an error.
NCTM Standards: Percentage Decrease, Sequences.

19. The Probability of the Perfect Skate Routine

Valid first trick: 5 options
Second trick:

- If first = flip (kickflip/heelflip): next cannot be flip → 3 choices
- If first ≠ flip (180, grind, wall ride): next can be any except same-type restriction only applies to flips → 5 choices

The third trick must be 180 spin, so 1 option.

Calculate valid sequences:

Case A – First trick is flip (2 options)
Second: 3 choices
Third: 1
Total: $2 \times 3 \times 1 = 6$

Case B – First trick is non-flip (3 options)
Second: 5 choices
Third: 1
Total: $3 \times 5 \times 1 = 15$

Total favorable = 21 sequences.

Total possible (without any restrictions): $5 \times 5 \times 5 = 125$
But restrictions remove sequences with consecutive flips.

Theoretical probability = 21 / (total valid sequences).
Total valid sequences = 125 – (all sequences containing consecutive flips).
(You may optionally compute explicitly.)

NCTM Standards: Statistics & Probability, Ratios

20. The Vanishing Drone Coordinates

Distance between (–3, 7) and (5, –1):

$\Delta x = 8$, $\Delta y = -8$

Distance = $\sqrt{(8^2 + (-8)^2)} = \sqrt{(64 + 64)} = \sqrt{128} = 8\sqrt{2}$ units.

NCTM Standards: *Geometry, Expressions & Equations*

Tools

Math Reasoning & Explanation Rubric & Reflection Guide

Name: ______________________________ **Date:** ____________________

Mystery/Problem Title: ____

Category	4 – Exceeds Expectations	3 – Meets Expectations	2 – Approaching Expectations	1 – Beginning
1. Understanding the Problem	Clearly identifies all important information, variables, and relationships; restates the problem in own words with full accuracy.	Identifies most important information and understands the main question.	Identifies some information but misses or misinterprets key details; may restate incorrectly.	Shows limited understanding; cannot identify what the problem is asking.
2. Strategy Selection & Planning	Chooses efficient, logical strategies; plans multi-step approach; may consider more than one method.	Chooses an appropriate strategy and follows a reasonable plan.	Strategy is partially effective but incomplete or inefficient; plan may be unclear.	No clear strategy; random attempts or guesswork.
3. Mathematical Reasoning & Accuracy	Uses reasoning that is clear, thorough, and mathematically sound; calculations are correct and justified; connected steps.	Reasoning is mostly clear; steps follow logically; minor calculation or reasoning errors may be present.	Some correct reasoning but includes major errors or missing steps; connections may be unclear.	Work contains little correct reasoning; major misunderstandings or inaccurate steps.

Category	4 – Exceeds Expectations	3 – Meets Expectations	2 – Approaching Expectations	1 – Beginning
4. Use of Representations (Models, Diagrams, Equations, Tables, Charts, etc.)	Uses multiple representations that enhance understanding; diagrams/models are accurate, labeled, and connected to solution.	Uses at least one appropriate representation to support the solution.	Representation is attempted but incomplete, inaccurate, or not well connected to reasoning.	No representation used, or representation is unrelated to the problem.
5. Communication & Explanation of Thinking	Explanation is thorough, organized, and shows clear justification for each step; uses math vocabulary accurately.	Explanation is understandable and includes the main steps; some justification provided.	Explanation is minimal, unclear, or missing reasoning; steps not well described.	No explanation or explanation does not match the work shown.
6. Final Answer & Check	Final answer is correct, clearly labeled, and includes evidence of checking work using a second method or reasoning.	Final answer is correct and labeled; checking may be simple or implied.	Answer is incomplete, incorrect, or unlabeled; no checking.	No final answer or answer is unrelated to the problem.
7. Persistence, Effort, and Problem-Solving Habits (Optional Category)	Shows perseverance, revision of ideas, and multiple attempts; uses feedback or self-correction effectively.	Completes the task with steady effort; may revise minor mistakes.	Needs prompting to continue; effort inconsistent.	Gives up easily, makes minimal attempt, or does not try.

Scoring Guides

You can use the rubric in several ways:

Total Points (if all 7 categories used):

- Maximum = 28 points
- 24–28 → Advanced
- 18–23 → Proficient
- 12–17 → Developing
- 7–11 → Beginning

Simplified 5-Category Version

If preferred, use only:

- Understanding
- Strategy
- Accuracy
- Representation
- Explanation

Student-Friendly Problem-Solving Checklist

___ I understood the problem and found the important information.

___ I made a plan and chose a good strategy.

___ I showed my work clearly and used math models when needed.

___ My answer makes sense and is labeled.

___ I explained my thinking in a way someone else can follow.

Sentence Starters to Help Me Explain My Thinking or Craft My Reflection

I can use these in my work or when talking about math:

- First, I…
- I chose this operation because…
- I solved it by…
- This model/diagram shows…
- Another way to solve this could be…
- My answer makes sense because…
- I checked my work by…

Teacher's Check List & Notes

Name: ______________________________ **Date:** ____________________

Mystery/Problem Title: ____

Strengths:

□ Clear steps

□ Good math vocabulary

□ Helpful models/pictures

□ Strategy makes sense

□ Answer is correct and checked

Next Steps:

□ Explain more clearly

□ Add math words

□ Add a picture/model/equation

□ Check my work

□ Try another strategy

Math Reasoning & Explanation Rubric & Reflection Guide for Groups

Group Name/Number: ___________________________

Mystery/Task: ___________________________________

Date: ________________ **Score:** 1 2 3 4 (circle one)

Score	Group Work Shows…
4 – Expert Problem Solvers	We explained our math thinking **clearly** and **worked together** to show every step. We used **math vocabulary** and at least one **diagram, table, model, or equation**. Our answer is **correct and makes sense**. We checked it and could **teach someone else how we solved it**.
3 – Strong Math Team	We explained **most** of our thinking. We used some math words and at least one picture, model, or equation. Our answer is correct and makes sense. We worked together well.
2 – Still Working	We explained **some** of our thinking, but parts are unclear or missing. We may not have used a helpful model or math words. There may be mistakes in our answer or steps.
1 – Getting Started	Our explanation is unclear, we did not show how we solved it, or our math does not match the problem. Little or no teamwork is shown. Our answer is incorrect or not supported.

Helpful Group Math Talk Starters

Groups should try to use statements like:

- I think the problem is asking us to…
- Let's try this strategy because…
- I noticed a pattern…
- Can someone explain why that step works?
- Is our answer reasonable? How do we know?
- Let's check our math another way.

Teacher's Check List & Notes

Group Name/Members/Number: ________________________________

Mystery/Task: ______________________________

Date: ______________

Strengths:

□ Clear steps

□ Good math vocabulary

□ Helpful models/pictures

□ Strategy makes sense

□ Answer is correct and checked

□ Effective group collaboration evident

Next Steps:

□ Explain more clearly

□ Add math words

□ Add a picture/model/equation

□ Check my work

□ Try another strategy

Teacher Resource Appendix

BRIGHT CHILD BOOKS, LLC
BrightChildAZ.com
BOOKS, PUBLISHING, & MORE

Teacher Resource Appendix

This appendix provides a quick overview of the **skills, standards, and mathematical practices** addressed in *Math Mysteries for Curious Minds.* Use it as a reference when planning lessons, aligning to curriculum standards, or differentiating for student needs.

I. Mathematical Habits of Mind

Each mystery is designed to strengthen these core habits of gifted mathematical thinkers:

Habit of Mind	Description	Example in Action
Perseverance	Tackles challenging problems with persistence and flexibility.	Continuing to test solutions even after early attempts fail.
Precision	Attends to detail, units, and accuracy in communication.	Labeling answers with correct units of measure.
Logical Reasoning	Builds arguments and identifies patterns or inconsistencies.	Eliminating impossible outcomes in a mystery using given clues.
Pattern Recognition	Notices structure and repetition in numbers or geometry.	Finding a rule behind a coded message or patterned sequence.
Creative Problem-Solving	Applies math in novel, imaginative contexts.	Using mathematical reasoning to solve a detective-style puzzle.
Communication	Explains reasoning verbally and in writing.	Presenting mathematical evidence to support a solution.
Reflection	Evaluates strategies, mistakes, and alternative solutions.	Writing in a math journal: "Next time, I would try…"

II. Integration with Gifted Standards

The mysteries also align with key **National Association for Gifted Children (NAGC)** program standards for advanced learners:

Gifted Standard	Connection to Math Mysteries
Critical Thinking & Problem Solving	Students analyze multi-step problems with multiple valid solution paths.
Creativity & Innovation	Learners are encouraged to create their own mysteries and strategies.
Depth & Complexity	Mysteries require connecting multiple concepts, not isolated skills.
Self-Directed Learning	Students make independent decisions on how to approach each case.
Communication & Collaboration	Group investigations emphasize explanation and respectful argumentation.

III. Sample Extension Questions for Deeper Learning

Use these prompts to increase rigor or extend a mystery beyond its base level:

- "What if one of the clues was wrong—how would it change your answer?"
- "Can you represent this mystery as a graph or table?"
- "How might this same situation work with larger numbers or decimals?"
- "Can you write a new version of this mystery that includes fractions or measurement conversions?"
- "What patterns or shortcuts could make your solution more efficient?"

These questions reinforce mathematical habits of inquiry and creativity that gifted students need to develop sustained, flexible thinking.

IV. Assessment & Reflection Tools

Suggested Rubric Categories (1–4 Scale):

- **Understanding of Problem:** Identifies relevant information and goal.
- **Mathematical Strategy:** Chooses efficient and logical approaches.
- **Accuracy:** Computes correctly, attends to precision.
- **Reasoning & Justification:** Explains thinking clearly and supports it with evidence.
- **Creativity & Insight:** Demonstrates original connections or alternative solutions.
- **Perseverance:** Demonstrates persistence through challenge or error.

Optional Student Reflection Prompts:

- "What was your biggest 'aha!' moment?"
- "How did you handle a mistake or dead end?"
- "What new strategy did you try today?"
- "What part of this mystery reminded you of real life?"

V. Final Note for Educators and Parents

When students engage with these mysteries, they're not only learning math—they're developing the habits of **mathematical thinkers, inventors, and detectives.**
Each case challenges them to balance logic and creativity, curiosity and precision, imagination and discipline.

Encourage them to **embrace the challenge** and **trust their reasoning**—because every great mathematical mystery begins with the words:

> "I wonder what would happen if…"

About the Author

Adam C. Laningham brings more than two decades of experience in public education and a deep passion for supporting gifted and twice-exceptional learners. Recognized as *Arizona Gifted Teacher of the Year*, Adam has taught across multiple grade levels, developed and led innovative gifted programs, and managed district-wide initiatives serving over 6,000 gifted students.

As founder of Bright Child Books, Adam is the author of several works dedicated to nurturing the potential of advanced learners. He also established The Gifted Collective, a thriving learning center that supports gifted and 2e children and their families.

Adam's leadership extends nationally and internationally. He served six years on the board of Supporting the Emotional Needs of the Gifted (SENG), including serving as President, and contributes his expertise as a member of the Arizona Association for Gifted and Talented (AAGT) Advisory Council, a founding member of Callisto (supporting gifted foster youth), and an advisor for CogAT Riverside Insights. Most recently, he founded the nonprofit Growing Intellect & Fostering Talent.

Recognized as an engaging speaker and trusted consultant, Adam is committed to creating opportunities, resources, and communities that empower gifted learners to thrive—academically, socially, and emotionally.

adam@brightchildbooks.com

Please contact us about workshops and training!

Further Resources

Get Your Kids Thinking!

Bright Child Books
PROFESSIONALLY CURATED RESOURCES FOR CURIOUS LEARNERS

Real world math that is fun, meaningful, and interesting!

www.BrightChildBooks.com

Books, Training, & More

Further Resources

Get Your Kids Thinking!

Bright Child Books
PROFESSIONALLY CURATED RESOURCES
FOR CURIOUS LEARNERS

www.BrightChildBooks.com

Books, Training, & More

Further Resources
Gifted, Twice-exceptional, & Trauma Support

Bright Child Books
PROFESSIONALLY CURATED RESOURCES FOR CURIOUS LEARNERS

www.BrightChildBooks.com

Books, Training, & More

BRIGHT CHILD BOOKS, LLC
BrightChildAZ.com
BOOKS, PUBLISHING, & MORE

Further Resources

Get Younger Kids Thinking!

Bright Child Books
PROFESSIONALLY CURATED RESOURCES FOR CURIOUS LEARNERS

PreK - 1st Grade
Animal Mysteries

Grades 2-3
Stories With Holes Junior

Gradws 3+
Stories With Holes

www.BrightChildBooks.com

Books, Training, & More

Further Resources

Get Your Kids Thinking!

Bright Child Books
PROFESSIONALLY CURATED RESOURCES
FOR CURIOUS LEARNERS

www.BrightChildBooks.com

Books, Training, & More

Thank you so much for purchasing one of our resources! We are continuously working on improving our titles, so we would love your input. If you see anything that needs to be adjusted or you have any recommendations, please email us.

adam@BrightChildBooks.com

Also, we would love it if you could take a minute and leave us a positive review on Amazon. Every little bit helps small publishers and authors get their work out there!

BRIGHT CHILD BOOKS, LLC
BrightChildAZ.com
BOOKS, PUBLISHING, & MORE

www.ingramcontent.com/pod-product-compliance
Ingram Content Group UK Ltd.
Pitfield, Milton Keynes, MK11 3LW, UK
UKHW051333111026
15762UKWH00030B/877